# YOUR KNOWLEDGE HAS VALUE

- We will publish your bachelor's and master's thesis, essays and papers

- Your own eBook and book - sold worldwide in all relevant shops

- Earn money with each sale

Upload your text at www.GRIN.com and publish for free

Sarah Schmidt

# Impacts of Mining Activities on Protected Areas and their Mitigation. Two Cases from Latin America

GRIN Publishing

**Imprint:**

Copyright © 2015 GRIN Verlag, Open Publishing GmbH
Print and binding: Books on Demand GmbH, Norderstedt Germany
ISBN: 978-3-668-01109-0

**This book at GRIN:**

http://www.grin.com/en/e-book/302514/impacts-of-mining-activities-on-protected-areas-and-their-mitigation-two

Humboldt Universität zu Berlin

Faculty of Life Sciences

Module: Biodiversity and Conservation Management

Winter Semester 2014-15

# Impacts of mining activities on protected areas, and their mitigation. Two cases from Latin America

Sarah Schmidt

2. FS M.Sc. Integrated Natural Resource Management

# Table of Contents



**Figures:**



**Tables:**

# 1. Introduction

Latin America is rich in vulnerable biodiversity of flora and fauna, outstanding landscapes and ecosystems and natural resources. Protected areas (PAs) are a tool for the conservation of land cover and ecosystem services (JOPPA; PFAFF, 2010, p. 1633). But the world demand for natural resources is rising and with it resource extraction and intensive land use (SWENSON et al, 2011, p. 1).

Metal mining activities are one of the biggest threats towards PAs and these activities are steadily increasing (DURÁN et al, 2013, p. 273). Deep environmental and social conflicts are associated with mining activities worldwide and especially in Latin America.

The research question of this paper is: What are potential social and environmental impacts of mining activities on PAs and how can they be mitigated?

The aim of the paper is to figure out important impacts of mining activities on PAs and to highlight possible solutions for mitigation.

In chapter two, which is the key chapter of this work, phases and forms of mining, environmental and social conflicts and the relevance for PAs, general solutions for mitigation and the current state of research will be pointed out.

Subsequently, in chapter three, the two case studies about Latin America, Intag cloud forest region in Ecuador and Huascarán National Park in Peru, will highlight the facts about mining in PAs of chapter two and present problems and solutions in a practical manner.

The last part of this work, chapter four, will conclude the facts which were figured out and compare the two practical case studies regarding to mining impacts and their mitigation.

# 2. Mining – An Overview

## 2. 1 Phases and Forms of mining and interconnected environmental impacts

There are different phases of a mining project and forms of mining and each phase of mining is linked to different environmental impacts (ELAW, 2010, p. 3).

As a first step, the **exploration** takes place, where information about the value and location of the mineral will be collected. This step includes surveys, field studies and the drilling of boreholes for testing purposes. In order to allow also the entrance of heavy vehicles, a clearing of big vegetation areas may be involved in this phase (ELAW, 2010, p. 3).

If it can be proven through the exploration phase, that there is a big mineral ore deposit, the **development phase** of the mine, which includes the construction of access roads, site preparations and clearing of land, will be initiated. These activities also may cause profound environmental impacts, particularly if they are taking place within or bordering ecologically sensitive areas. (ELAW, 2010, pp. 3+4).

According to the ELAW (2010, p. 4) once the exploration and development phases of the mining process are accomplished, the **active mining** which in other words means: extraction and beneficiation of metals from the earth, can take place. The most common forms of mining will be listed and shortly explained below:

- ***Open-pit mining****: this method is one of the most environmentally destructive forms of mining, particularly in tropical forests, due to the removal of natively vegetated zones in form of tree logging, clear-cutting or burning of vegetation. It is a type of strip mining in which it is necessary to remove layer upon layer of overburden and ore because the ore deposit extends really deep into the ground.*

- ***Placer mining****: also called 'hydraulic mining' because the metal of interest, which shall be removed, usually gold, is located into a streambed or floodplain. This method is also environmentally destructive because it releases large quantities of sediment, which may have an impact to surface water for many miles downstream of the mine.*

- ***Underground mining:*** *using this form of mining, a small amount of overburden is removed in order to gain access to the ore deposit through tunnels or shafts. It is on the one hand less environmentally destructive but on the other hand more expensive and includes higher safety risks than open-pit mining. Many large underground mines are operating around the whole world.*

- ***Reworking of inactive or abandoned mines***: *this form of mining refers to the reworking of waste piles (tailings) from inactive or abandoned mines by re-extracting metals from mining waste. Here the environmental impacts of open-pit mining and placer mining can be avoided but there will be still environmental impacts due to the purification of metals from the waste.* (ELAW, 2010, p. 4)

As per ELAW (2010, pp. 5+6), after and while the active mining takes place, **overburden and waste rock**, which usually has an enormous quantity, **have to be removed** in order to reach the metallic ores. Sometimes, these wastes contain alarming levels of toxics, for example cadmium or arsenic. Afterwards the **ore extraction** phase takes place with the help of heavy equipment and machinery, which again causes environmental impacts, such as dust emissions. During the **beneficiation** phase the ore will be grinded and the metal will be separated from the non-metallic ore material. Hereby 'tailings', which is high-volume waste, will be generated. Hence, one of the central questions, whether a mining project is environmentally acceptable or not is: How does a mining company dispose the toxic waste material? The long-term goal of the disposal of tailings is to prevent the release of toxic parts into the environment (ELAW, 2010, pp. 5+6).

According to ELAW (2010, p. 7), the last phase of mining might be the **closure of the site**, which sounds rather easy, but many factors have to be taken into account in order to create the condition of 'pre-mining'. Questions like: how will the release of toxic substances from the mining facilities continuously prevented and financed have to be carefully answered (ELAW, 2010, p. 7).

## 2. 2 Environmental and social conflicts in connection with mining and the relevance for protected areas (PAs)

According to DURÁN et al (2013, pp. 272-277) there is currently both increasing demand and prices for metals worldwide, which increases consequently metal mining activities especially into more remote and currently un-mined areas, often in or close to PAs, which play a major role in biological conservation. According to SONTER et al (2013, p. 6300) the establishment and operation of a mine involves, among others, deforestation, soil displacement, water and energy consumption and development of large industrial infrastructure which have potentially negative environmental effects of biodiversity and the service and function of ecosystems (SONTER et al, 2013, p. 6300). Actually, mining companies cause the most profound and mostly irreversible harms to the natural environment in comparison with other industrial sectors and furthermore have many negative social impacts, inclusively industrial accidents, health and safety issues, among others (MUTTI et al, 2012, p. 212).

Around 7 % of mines for the metals aluminum, copper, iron and zinc have a direct overlap with PAs and 27 % lie within a 10 km area of a PA boundary. Mining activities can affect the environment over big distances with a long durability (even after the mining site was closed, as mentioned before). Hence there is an urgent need of restricting or mitigating these conflicts in order to stabilize or raise the conservation performance of PAs and its biodiversity (DURÁN et al, 2013, p. 272).

Some of these environmental and social conflicts, which can come up will be listed below:

**Environmental conflicts:**

**Water resources**: According to ELAW (2010, pp. 8-12) one of a leading impact of mining is the effect on surface- and groundwater quality and water resource availability within the mining area. The quality and availability of water should be secured, on the one hand for human consumption and on the other hand for aquatic life and terrestrial wildlife. One of the most serious threats to the water resources is the acid mine drainage formed by sulfides inside the rocks that are exposed due to metal mining. These sulfides can run uncontrolled into surface water or leach into groundwater.

**Air quality**: Pollutants, which occur during all phases of mining, especially during exploration, development, construction and operational phases can seriously harm people's health and the environment. These small pollution particles are easily widespread through the wind. In Figure 1 you can easily see how the emissions enter the atmosphere, undergo physical and chemical changes and last but not least impact the human health, the environment, the global climate, etc. (ELAW, 2010, pp. 8-12).

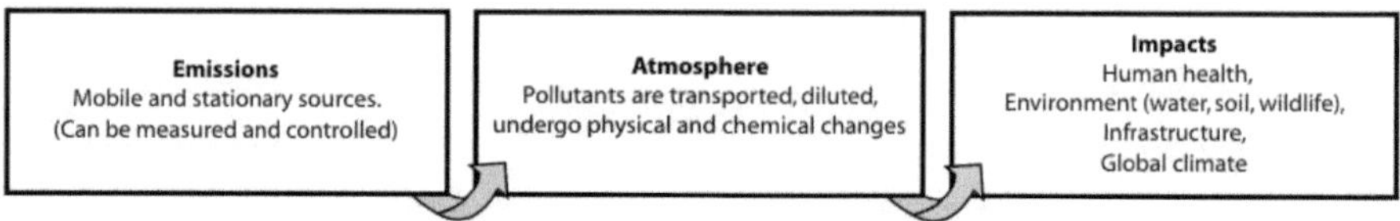


**Figure 1: Connection of Emissions, Atmosphere and Impacts**

Source: ELAW, 2010, p 12.

**Impacts on flora and fauna:** Flora and Fauna is endangered through the removal of vegetation and topsoil, the expulsion of fauna, noise generation and release of pollutants due to a mining project (ELAW, 2010, pp. 13-14). Table 1 shows probable effects of mining activities on the environment.

## Table 1: Probable effects of mining activities on the environment

Summary of examples of published studies that evaluate the extent of impact of mining activities on various ecological and environmental variables.

| Authors | Mine type | Ecological/environmental effect | Maximum distance impact from mining source (km) |
| --- | --- | --- | --- |
| Hernández et al. (1999) | Pyrite | Bird mortality | 25 |
| Vásquez et al. (1999) | Copper | Macroalgae abundance | 3 |
| Razo et al. (2004) | Copper–gold, lead–zinc–silver | Heavy metal concentration | 5 |
| Telmer et al. (2006) | Copper | Heavy metal concentration in lake sediments | 50 |
| Yakovlev et al. (2008) | Nickel | Soil quality | 25 |
| Kodirov and Shukurov (2009) | Copper and zinc | Heavy metal concentration in soil | 4 |
| Kuznetsova (2009) | Copper | Collembola communities in coniferous forests | 7 |
| Lafabrie et al. (2009) | Cobalt | Heavy metal concentration in seagrass | 5 |
| Taylor et al. (2009) | Copper, zinc and lead | Downstream water quality | 30 |
| Bonifait and Villard (2010) | Peat | Odonate abundance | 1 |
| Chauhan (2010) | Zinc | Deforestation | 11[a] |
| Huang et al. (2010) | Copper and zinc | Water acidity and heavy metal concentration | 10 |
| Katpatal and Patil (2010) | Coal | Flooding | 15 |
| Lefcort et al. (2010) | Copper and zinc | Stream insect diversity and abundance | 2.5[a] |
| Vodyanitskii et al. (2011) | Copper | Decrease of soil quality | 30 |

[a] Distance impact was calculated as the perimeter of a circle with stated area mining impact.

Source: DURÁN et al, 2013, p. 276

## Social conflicts:

According to ELAW (2010, p. 15) the social impacts of large-scale mining projects are complex and debatable. One the one hand, mining projects may create jobs, a better infrastructure and increase the demand of goods and services in remotely located areas (ELAW, 2010, p. 15). But on the other hand, mining companies cause negative impacts on livelihoods of local communities and offences of human rights (MUTTI et al, 2012, p. 212).

If local communities feel inadequately treated or uncompensated, mining projects can even lead to violent conflicts (ELAW, 2010, p. 15).

**Displacement and resettlement:** Displacements and resettlements of communities are common social impacts of mining and cause a lot of conflicts. Communities are not only losing their home, but also their land and hence their livelihoods (ELAW, 2010, p. 15).

**No access to clean water:** As mentioned before one of a leading impact of mining is the effect on surface- and groundwater quality as well as water availability within the mining area (ELAW, 2010, p. 8). Contaminated ground- and surface water can harm the health of the people living in the communities close to mining areas. Due to that, there are many conflicts between the miners and the communities (ELAW, 2010, p. 16).

**Effects on livelihoods:** All environmental conflicts due to mining projects (some of them have been mentioned before) can be transferred to livelihoods. According to this they are endangered through air pollution, water pollution and degradation, soil pollution, effects on flora and fauna, effects on biodiversity, etc. (ELAW, 2010, p. 16).

## 2. 3    General solutions and regulations for mitigation

Now we can see more clearly what mining means and how mining projects may impact the natural environment and the society. Hence we may now know, that it is very important to regulate mining activities in a reasonable manner, including the conservation of PAs which are important for many reasons. According to PHILLIPS (2001, p. 6) PAs are important for example because of biodiversity conservation, homelands for local communities, provision of natural products, protection of watersheds and soils, recreation and tourism, etc. (PHILLIPS, 2001, p. 6). Some of the possibilities in this regard will be discussed below.

First of all, to know about the possible impacts of a mining project before it starts at all, it would be highly recommended to conduct an environmental impact assessment (EIA). The EIA should be compulsory for every mining project. According to ELAW (2010, p. 19) using these EIA guidelines, will ensure, that environmental considerations are involved into a decision making about setting up a mining project and how possible impacts could be mitigated. Furthermore the public will be informed about environmental consequences (ELAW, 2010, p. 19) and hence the whole process will be more transparent. According to ELAW (2010, p. 67) the whole mining process also needs to be monitored through the mining company and/or government officials to measure its impacts on the environment (ELAW, 2010, p. 67) and respond directly to the concerns identified by the EIA process (ROSENFELD; CLARK, 2000, p. 41).

Moreover there are different measures of mitigation and examples of action as it can be seen in Table 2 (DURÁN et al, 2013, p. 277).

**Table 2: Mitigation measures and examples of action**

| Mitigation measure | Example of action |
| --- | --- |
| Avoidance | To avoid infrastructure in priority areas for biodiversity using spatial planning methods |
| Minimization | Establishment of ecological corridor and buffer zones |
| Restoration | To restore connectivity between patches of habitats within landscapes. Reforestation |
| Offset | Environmental compensation policies and payment of ecosystem services schemes |

Source: DURÁN et al, 2013, p. 277

It can be recognized, that the mitigation measures mentioned in the table above are hierarchical ordered. Avoidance is the strongest and offset the weakest measure of mitigation. Herewith is meant that what cannot be avoided should be at least minimized, taking into account for restoration or offsetting. These measures should be already taken into account during the planning of a mining project (DURÁN et al, 2013, p. 277) and should be as well compulsory.

Furthermore, according to ELAW (2010, p. 16) it is important that proponents of mining projects are insuring that the basic rights of individuals or communities, for example the right to clean water, are not expulsed. Such rights should be tightened in national laws, based on international human rights (ELAW, 2010, p. 16).

Generally, according to ROSENFELD; CLARK (2000, p. 5), extensive social, environmental and mining legislations and regulations should be determined. They need to be clearly and transparently formulated. For example which parties do have responsibility for which issues as well as monitoring and enforcement of legislations and regulations, in order ensure an effective realization and to give room for enhancements (ROSENFELD; CLARK, 2000, p. 5). An example is the use of „environmental compensation policies", as per SONTER et al (2013, p. 6301), where mining companies legally need to `compensate´ for their mining activities and hence impacts to the environment.

In order to make companies comply with these social and environmental regulations regarding mining activities, governments could also use financial and economic incentives (ROSENFELD; CLARK, 2000, p.5). These incentives may include performance bonds, trust funds or offsets.

Another promising strategy for the regulation of mining activities could be long-term land use-plans, according to ROSENFELD; CLARK (2000, p. 5). Governments could develop under close cooperation with communities, companies and/or other stakeholder's land-use plans, which consider ecological, geological and cultural priorities for the entire country or specific regions. Here should be also included the designation of areas which are too sensitive in regard to the environment or the culture, to allow mining activities (ROSENFELD; CLARK, 2000, p. 5).


## 2. 4    Current state of research

Currently, there is a wide range of discussions about several topics regarding mining and its impacts on PAs inclusive their mitigation going on. Some of the most relevant findings for this paper will be presented here.

According to DURÁN et al (2013, p. 272) PAs are highly needed for biodiversity conservation and buffering it from external threats. PAs are more successful in fulfilling those functions compared to non-protected areas. But nevertheless there are numerous ways in which PAs regarding to structure, management, distribution and extent still need to be improved. As per DURÁN et al and many other researchers, one of the biggest threats nowadays towards PAs are metal mining activities. These activities have increased tremendously from 1992 to 2010, what can be seen in the graph below (DURÁN et al, 2013, p. 273).

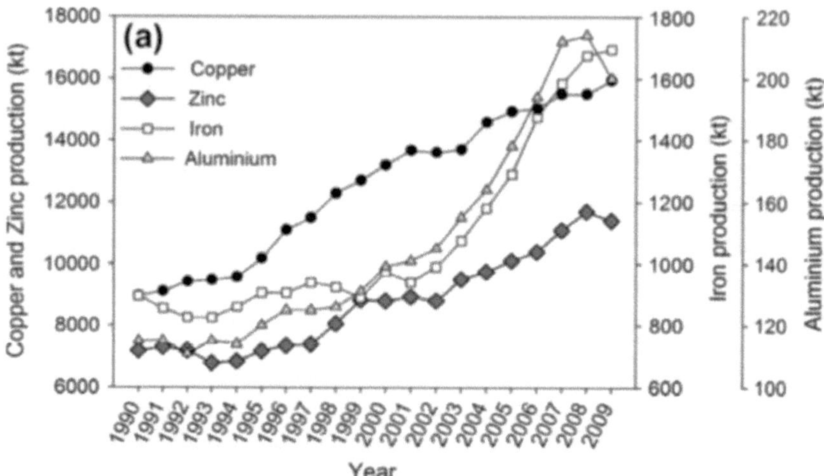


**Figure 2: Global production of aluminum, copper, zinc and iron from 1990 to 2010**

Source: DURÁN et al (2013, p. 273)

There is a higher than expected proportion of mining activities in or close to PAs which affect the conservation performance of many PAs (DURÁN et al, 2013, p. 277). Previous studies which were conducted in alpine and cloud forest rivers in Latin America point relations between land use and water quality and especially mining activities are in relation with lower pH values, increased concentrations of arsenic, copper, zinc, cadmium and lead and a reduced aquatic invertebrates biodiversity (KNEE; ENCALADA, 2013, p. 387). This finding is especially important for the case study about Ecuador in Chapter 3.

Mining for aluminum, iron, zinc and copper are widely distributed across the world with higher concentrations in the Andes mountain range, among others. The following maps show the global mining distributions of bauxite, copper, zinc and iron, the distribution of PAs and PAs with mining activities at different proximity levels (DURÁN et al, 2013, p. 274).

3.1.2 Mining activities within the Intag region and related environmental and social impacts

According to KNEE; ENCALADA (2013, pp. 385-386), mining companies focus since the early 1990s on Intag´s rich mineral resources. Between 1991 and 1995, a subsidiary of Mitsubishi drilled 24 mining boreholes in the area of Junín, a known deposit of Intag. But due to strong community resistance, these boreholes were later abandoned. Currently, only a small goldmine in the region is operating. But nevertheless, due to high prices of metals like copper and gold, as mentioned in chapter 2.2, the pressure of mining activities will probably continue in the region. Hence, the threat of open-pit copper mining and related big environmental concerns will remain (KNEE; ENCALADA, 2013, pp. 385-386). The local grassroots environmental organization DECOIN (Intag Defense and Ecological Conservation) was founded in 1995 with the aim to fight against the big mining companies in the Intag region. According to DECOIN (2014), a mining project would be located in immaculate cloud forests, threatening, as per an environmental impact study, the forests, rivers and communities through relocations as well as contamination with heavy metals. In addition, the habitats of 28 species of endangered mammals and birds would be threatened (DECOIN, 2014).

The maintenance of good water quality within the Intag region is very important for the biodiversity and human health. But according to the studies of KNEE; ENCALADA (2013, p. 396) the highest dissolved metal concentrations were detected in sites with mining activities as well as in the abandoned boreholes in Junín. That means, that mining as a land use could have the most serious impacts on water quality in the Intag region. However, the dissolved metal concentrations in the Junín streams were higher compared to those concentrations in other Intag streams but still lower than in other areas with more extensive mining activities in Ecuador or Bolivia. Hence it´s supposed that metal concentrations could rise to harmful levels, if mining activities in Intag will increase (KNEE; ENCALADA, 2013, pp. 396-397).

3.1.3 Means of mitigation

The Junín mining project is strongly rejected by most of the Intag communities (DECOIN, 2014). The threat of open-pit copper mining has induced the local population, to build up a strong community organization (KNEE; ENCALADA, 2013, p. 386). As per DECOIN (2014), it is anyways not possible to guarantee in developing countries, like Ecuador is, a long-term conservation of natural territories without a local community participation.

Furthermore, in contrast to chapter 2.4, local governments with the support of Cotacachi Municipal Government are actively against the mining projects. Due to the rejections of the mining project and civil society pressure, the government took away the permissions for companies to start mining activities in the Intag region in late 2008 (DECOIN, 2014).

According to DECOIN (2014), as an alternative to mining activities, an ecological tourism site has been established in the Intag region in 2002. The business is wholly managed by around 30 members oft he community. Some of the profit goes to a special development fund for the community.


## 3.2 The case of the Huascarán National Park in Peru

3.2.1 Overview of the Huascarán National Park

According to SHOOBRIDGE (2005, p. 1-3), the Huascarán National Park comprises 340,000 hectares and was created in 1975. It is an internationally recognized UNESCO Biosphere Reserve and on the World Heritage List since 1985. The park is located in north-central Peru in the department of Ancash and is ecologically part of the Puna biogeographic province. It includes almost the whole Cordillera Blanca Mountain Range inclusive snow-covered peaks of altitudes between 5000 and 6768 m.

The three main objectives of the park are:

- conservation of outstanding biodiversity, including endangered and rare species of flora and fauna

- protection of the landscape quality inclusive ecosystem quality

- protection of the water quality and quantity including the stability of the hydrologic cycle. People are depending on the Cordillera Blanca watersheds.


The Huascarán National Park faces a lot of threats, among others loss of vegetative coverage, tourism, global warming and mining. It is hence in a vulnerable situation (SHOOBRIDGE, 2005, p. 1-3).

3.2.2 Mining activities within the Huascarán National Park and related environmental and social impacts

As per SHOOBRIDGE (2005, pp. 40-42), mining activities within a national park are in general legally not allowed. But in this case, mining authorizations have been issued already before the region was declared Huascarán National Park. Mining companies have got rights for exploitation as long as they take into account current environmental regulations. There are 78 mining admissions within the park and 9 are at present working, mostly situated in the southern part of the park.

A big problem with mining within the area is that small mines are failing to comply with environmental regulations. For example the disposal of waste into water streams let tailings accumulate and acidic water conditions emerge.

Furthermore, abandoned, declining smelting operations have been detected at the bottom of Tuco Valley as well as abandoned mining tailings along the PAs southern edge. This represents a threat to water and soil and hence to the ecosystem in general.

Hence mining activities negatively influence the park objectives of landscape conservation, ensuring of water quality and biological diversity. Among others mining activities acidify water bodies, reduce vegetative coverage, bothers wildlife (noise, etc.) and mining waste accumulate and therefore degrading quality of landscapes.

Social conflicts include conflicts between communities living in the highlands of the park and mining companies, which change the areas ecology through mining activities and roads. People claim that the dialogue is often not informative or transparent. Furthermore, local people have high expectations towards mining companies about job opportunities and development. If this does not happen, expectations are broken and conflicts come up.

Additionally to the mining activities, rock and gravel extractions take place within the park. These extractions are responsible for negative impacts on landscape quality, soil stability, vegetation and they produce dust and introduce moreover heavy machinery (SHOOBRIDGE, 2005, pp. 40-42).

3.2.3 Means of mitigation

According to SHOOBRIDGE (2005, p. 54-55), the impacts of mining activities on soil and water bodies must be reduced. We already know that in the case of Huascarán National Park mining rights were issued before the area was declared a National Park. Nevertheless, mining activities should now be discouraged since they have huge impacts on ecological systems and social aspects. This could be reached with the application of mining norms and regulations with respect to PAs (SHOOBRIDGE, 2005, p. 54-55).

According to SHOOBRIDGE (2005, p. 41), environmental strategies would help to improve the mining´s sustainability within the Huascarán National Park. But so far they are not available. The Environmental Adjustment and Management Program (PAMA) would be needed to monitor currently active mines in order to guarantee compliance with environmental standards (SHOOBRIDGE , 2005, p. 41).

Furthermore, all mining operations, including those in buffer zones, should be more transparent and part of the Huascarán Work Group. Information about progresses in applying environmental management systems should be always provided. Mining companies and the Ministry of Energy and Mines should provide local authorities with results of mining controls and monitoring about liquid effluents and impacts of water bodies. Using this strategy, involved sectors and stakeholders will be always informed and exercise pressure towards mining companies, if they fail to comply with environmental standards and limits (SHOOBRIDGE , 2005, p. 55).


## 4. Resume and Conclusion

Mining activities can have profound environmental and social impacts on PAs. Especially in Latin America with it´s rich natural resources, biodiversity of flora and fauna and local communities that use these resources, many cases of conflicts can be found.

Comparing the two cases of Latin America, the Intag cloud forest region in Ecuador and the Huascarán National Park in Peru, it can be said that the cases are similar regarding to outstanding flora and fauna, including rare species, subsoil mineral resources and landscapes and ecosystems, including water quality and quantity, that need to be protected. Local communities that depend on these resources, for example on water resources for drinking and cooking, livestock and irrigation, are living in both regions.

In the case of Ecuador abandoned mines still negatively influence the water quality of the streams in the region. Also in the case of Peru, abandoned mines still contaminate water bodies and soil. In Peru active mines are still operating, because companies have got mining authorizations, before the area was declared a National Park.

The difference between the case of Ecuador and Peru is, that in Ecuador the PA is a self-declared 'environmental canton'. The communal government is actively opposing mining and supporting sustainable development. The community is strong and they were able to reject further mining activities. In the case of Peru, the PA is a declared UNESCO Biosphere Reserve and a World Heritage Site.

Mining would in both cases threaten the forest, rivers, habitats of animals, vegetation, communities due to change of ecological systems and road establishments. In Peru some companies fail to comply with environmental regulations and the procedures are often intransparent.

Although the Intag cloud forest region in Ecuador is not a declared PA like the Huascarán National Park in Peru is, mining has been so far successfully opposed due to a strong community organization as well as local and municipal governments that reject mining activities. They took away mining permissions of companies. Here can be seen that a strong local community can oppose mining activities better than an officially declared PA could do. In the case of Peru, governments could also take away the mining authorizations of companies, even though the National Park was declared later in that area. Now it is a National Park and that need to be respected. Moreover, governments need to at least develop mining norms and regulations as well as effective monitoring of mining activities, if not all mining rights could be taken away from companies.

Nevertheless, the economical factor of mining should also be taken into account. Jobs could be created and so the income of local communities will be ensured. But there are other opportunities. For example in the Intag cloud forest region, an eco-tourism site has been established by the community. Part of the generated money goes to a development fund for the communities. This alternative land use could also be taken into consideration by the government and management of the Huascarán National Park.

# References


DECOIN (2014): retrieved 05.12.2014 from http://www.decoin.org/accomplishments/ .

DURÁN , América P.; RAUCH , Jason; GASTON, Kevin J. (2013): Global spatial coincidence between protected areas and metal mining activities. In: *Biological Conservation*, Vol. 160, April 2013, pp. 272–278.

Environmental Law Alliance Worldwide (ELAW) (2010): *Guidebook for Evaluating Mining Projects EIAs*. Environmental Law Alliance Worldwide, Eugene (USA): 1st Edition.

JOPPA, Lucas N.; PFAFF, Alexander (2010): Global protected area impacts. In: *The Royal Society* Vol. 278, 2010, pp. 1633-1638.

KNEE, K. L.; ENCALADA, A. C. (2013): Land use and water quality in a rural cloud forest region (Intag, Ecuador). In: *River Research and Applications* Vol. 30, 2014, pp. 385–401.

MUTTI, Diana; et al (2012): Corporate social responsibility in the mining industry: Perspectives from stakeholder groups in Argentina. In: *Resources Policy*, Vol. 37, 2012, pp. 212-222.

PHILLIPS, Adrian (2001): Mining and Protected Areas. In: *Mining, Minerals and Sustainable Development*. London (UK): International Institute for Environment and Development (IIED), Nr. 62, October 2001.

ROSENFELD Sweeting, Amy; CLARK, Andrea P. (2000): *Lightening the Lode - A Guide to Responsible Large-scale Mining*. Conservation International. CI Policy Papers, Washington D.C. (USA).

SHOOBRIDGE, Diego (2005): *Huascarán National Park,* ParksWatch, Protected Area Profile – Peru, November 2005.

SONTER, Laura Jane; et al (2013): Mining, deforestation and conservation opportunities: A case study of the Quadrilátero Ferrífero land use change dynamics. In: EPIPHANIO, J. C. N.; GALVAO, L. S. (eds.): *Simpósio Brasileiro de Sensoriamento Remoto, 16. (SBSR)*. São José dos Campos (Brazil): INPE, April 2013, pp. 6300-6307. Available from: <http://urlib.net/3ERPFQRTRW34M/3E7G88S>. Access in: 2014, Nov. 28.

SWENSON, Jennifer J.; CARTER, Catherine E.; DOMEC, Jean-Christophe; DELGADO, Cesar I. (2011) : Gold Mining in the Peruvian Amazon: Global Prices, Deforestation, and Mercury Imports. PLoS ONE. Vol. 6, Issue 4, April 2011.